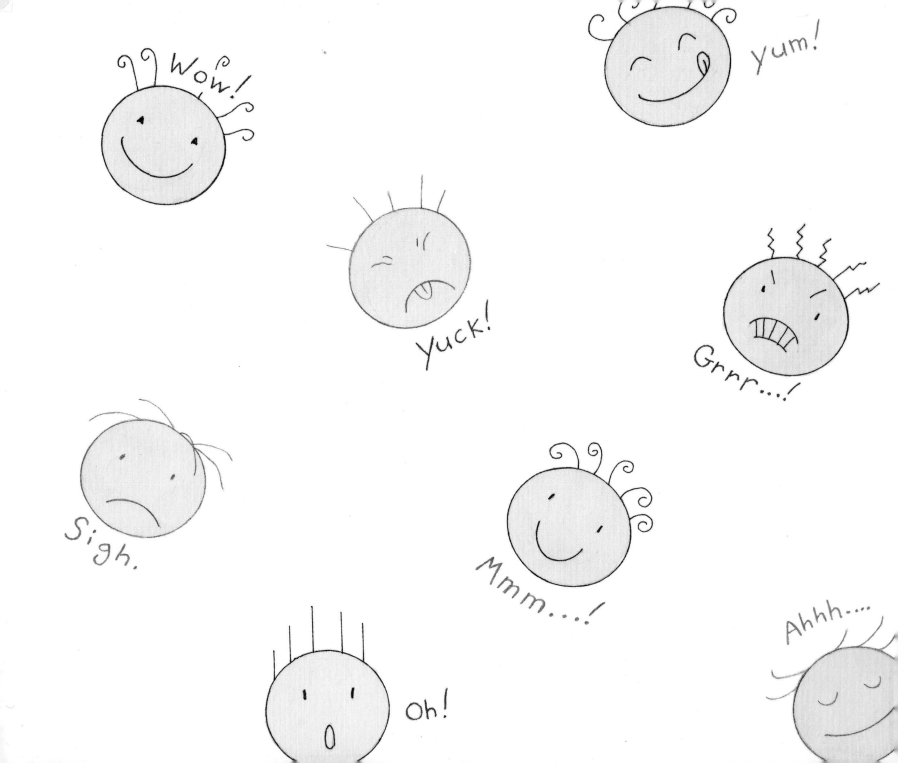

MathStart®
PROBABILITY

probably Pistachio

by Stuart J. Murphy • illustrated by Marsha Winborn

HarperCollins Publishers

LEVEL
2

In memory of M.E.M. and our many
urgent childhood stops for pistachio ice cream.
—S.J.M.

The publisher and author would like to thank teachers Patricia Chase,
Phyllis Goldman, and Patrick Hopfensperger for their help
in making the math in MathStart just right for kids.

HarperCollins®, 🐛®, and MathStart® are registered trademarks of HarperCollins Publishers.
For more information about the MathStart series, write to HarperCollins Children's Books,
10 East 53rd Street, New York, NY 10022 or visit our website at www.mathstartbooks.com.

Bugs incorporated in the MathStart series design were painted by Jon Buller.

Library of Congress Cataloging-in-Publication Data
Murphy, Stuart J., 1942–
   Probability: Probably pistachio / by Stuart J. Murphy ; illustrated by Marsha Winborn.
     p.   cm. – (MathStart)
    "Level 2."
   Summary: Readers are introduced to the concept of probability in a story about a boy who has a day
in which nothing goes right.
    ISBN 0-06-028028-X. – ISBN 0-06-028029-8 (lib. bdg.).– ISBN 0-06-446734-1 (pbk.)
    1. Probabilities–Juvenile literature.   [1. Probabilities.]   I. Winborn, Marsha, ill.   II. Title.   III.
Title: Probably pistachio.   IV. Series.
QA273.16.M87   2001   99-27695
519.2–dc21   CIP
                                                   AC

Typography by Elynn Cohen
12 13 SCP 20 19 18 17 16 15
❖
First Edition

It was Monday, and nothing was going right.
My alarm clock didn't go off.

I couldn't find my
favorite sneakers.

4

I tripped over my dog, Pirate,
and banged my knee.

And it was Dad's turn to fix lunch.

With Mom, Mondays mean it's probably pastrami.
And pastrami is my favorite food in the whole world.
But with Dad, you never know. It could be ham
and cheese. It could be peanut butter and jelly.
It could even be tuna fish.

I hate tuna fish.

6

7

When I got to school, all I could think about was pastrami. Then I remembered that Emma has a pastrami sandwich just about every day. Last week she had pastrami every day but Thursday.

In math class, I was so busy thinking about Emma's sandwiches, I didn't hear the teacher call my name.

And then I had to copy my homework over again because it was too soggy.

By the time I made it to lunch, there was only one seat left, right next to Emma. Maybe my luck was changing! "Want to trade lunches?" I asked. "I've got tuna fish."

"Great!" Emma said right away. "I love tuna fish."

After school, I went to soccer practice with my best friend, Alex.
I love soccer, so I didn't think anything could go wrong.

12

Coach usually has us count off like this: "One, two, one, two!" Then we split up into two teams and play a practice game. Alex and I always make sure there's somebody standing between us in line so we'll be on the same team.

Then Coach said, "We're going to try something new today, kids. Let's count off by threes so we'll have three small groups to practice kicking and passing."

I tried to switch with Chris so I'd still be on Alex's team.
But it was too late.

After practice, Coach always gives us a snack. Today he dumped some bags of pretzels and crackers into a big basket. He put in a few bags of popcorn too. Next to pastrami, popcorn is my favorite food in the whole world.

Coach held the basket up high while we all took turns picking. I saw Alex grab a bag of popcorn. I hoped there might be enough popcorn left for me.

16

When I got to the front of the line, I was still hoping for popcorn. "Hurry up and pick, Jack," Coach said.

I quickly grabbed a bag.

It was pretzels! I couldn't believe it. Next to tuna fish and liverwurst, pretzels are my least favorite food.

By the time I got home, I was starving. I mean, I'd
hardly had any breakfast because I'd been rushing
around. I sure hadn't eaten Emma's liverwurst sandwich.
And I saved my bag of pretzels for Pirate.

So when I walked in and the whole house smelled like pizza, I got really excited. Next to pastrami and popcorn, pizza is my favorite food in the whole world.

Dad was standing in the kitchen, stirring something in a big pot on the stove.

"Are we having pizza?" I asked.

"You know it's not pizza night," Dad answered. "I'm making spaghetti and meatballs."

23

Mom rushed in while I was setting the table. "I have a surprise!" she said. "I stopped at Frosty's on my way home from work."

"Ice cream for dessert!" Rebecca yelled. "Way to go, Mom."

"I got your favorite," Mom announced.

"Whose favorite?" I asked suspiciously. But she didn't hear me. Rebecca's favorite is chocolate. Now, everybody knows what my favorite is. I mean, next to pastrami and popcorn and pizza, pistachio ice cream is my favorite food in the whole world.

Mom reached in and pulled out a tub of ice cream.

"Thanks, Mom." Rebecca grinned. "You're the greatest."

I couldn't wait for this terrible day to be over.

"Hey! What's this?" Mom said. She reached into the bag again. "I think there's another tub in here. What do you suppose it could be?"

29

And this time I was right!

Finally things were looking up.

Maybe tomorrow I'd even get pastrami for lunch.

In *Probably Pistachio*, the math concept is probability: predicting the likelihood of any given event. Learning to make astute predictions helps children analyze data in order to make informed decisions.

If you would like to have more fun with the math concepts presented in *Probably Pistachio*, here are a few suggestions:

- Read the story together and ask the child to predict what he or she thinks will happen and why. Ask questions like: "Do you think Emma will have pastrami for lunch? Why do you think that?" As the child's understanding of probability grows, ask questions like: "Why didn't Jack's prediction come true?" "What question could Jack have asked Emma so that he might have made a better prediction?"

- After reading the story, ask questions like: "If Emma had a pastrami sandwich only once a week, would Jack expect pastrami when he traded with her?"

- Have the child keep track of what is served for school lunches for one week and then predict what will be served the next week.

- Ask your child to decide if certain events are likely, possible, or unlikely. Suggest events like: "You will go to bed at 8:30 tonight." "We will all go swimming this Saturday." "No one in your class will be absent tomorrow."

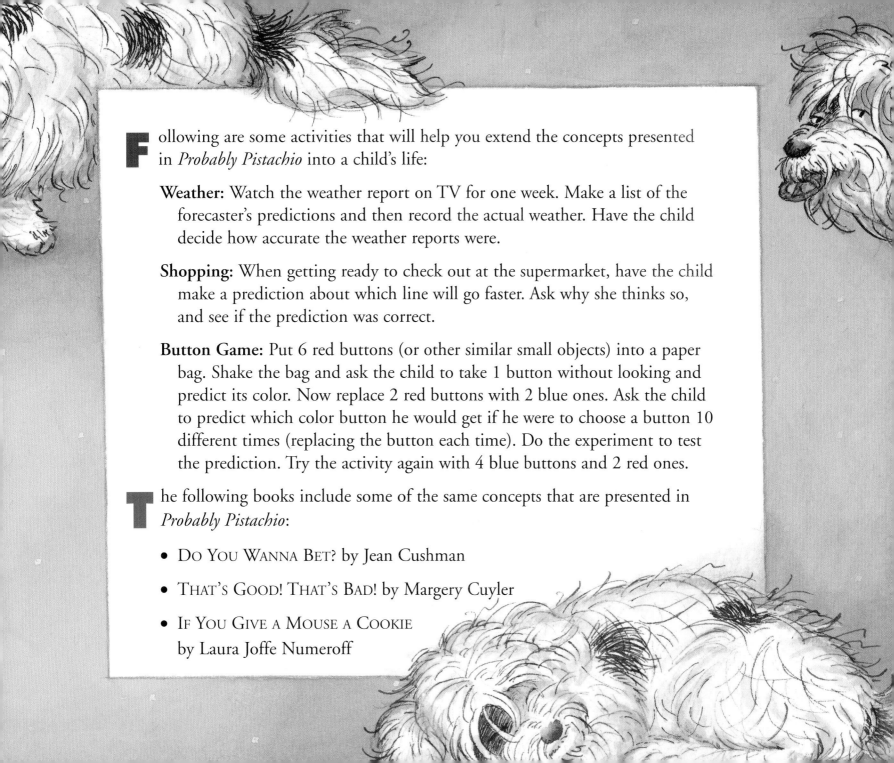

Following are some activities that will help you extend the concepts presented in *Probably Pistachio* into a child's life:

Weather: Watch the weather report on TV for one week. Make a list of the forecaster's predictions and then record the actual weather. Have the child decide how accurate the weather reports were.

Shopping: When getting ready to check out at the supermarket, have the child make a prediction about which line will go faster. Ask why she thinks so, and see if the prediction was correct.

Button Game: Put 6 red buttons (or other similar small objects) into a paper bag. Shake the bag and ask the child to take 1 button without looking and predict its color. Now replace 2 red buttons with 2 blue ones. Ask the child to predict which color button he would get if he were to choose a button 10 different times (replacing the button each time). Do the experiment to test the prediction. Try the activity again with 4 blue buttons and 2 red ones.

The following books include some of the same concepts that are presented in *Probably Pistachio*:

- Do You Wanna Bet? by Jean Cushman

- That's Good! That's Bad! by Margery Cuyler

- If You Give a Mouse a Cookie by Laura Joffe Numeroff